Count Your Way through

Kenya

by **Jim Haskins** and **Kathleen Benson**

illustrations by **Lyne Lévêque**

M Millbrook Press / Minneapolis

To Nigel, Nadia, and Nyon Cathéy –K. B.

To Samuel, Léa, and Renaud –L. L.

Text copyright © 2007 by Jim Haskins and Kathleen Benson
Illustrations copyright © 2007 by Millbrook Press, Inc.

Millbrook Press, Inc.
A division of Lerner Publishing Group
241 First Avenue North
Minneapolis, Minnesota 55401 U.S.A.

Website address: www.lernerbooks.com

Library of Congress Cataloging-in-Publication Data

Haskins, James, 1941–
    Count your way through Kenya / by Jim Haskins and
    Kathleen Benson ; illustrations by Lyne Lévêque.
       p.   cm. — (Count your way)
    ISBN-13: 978-1-57505-884-9 (lib. bdg. : alk. paper)
    ISBN-10: 1-57505-884-7 (lib. bdg. : alk. paper)
1. Kenya—Juvenile literature.   2. Swahili language—
Numerals—Juvenile literature.   3. Counting—Juvenile
literature. I. Benson, Kathleen. II. Lévêque, Lyne, ill. III. Title.
DT433.522.H37 2007
967.62—dc22   2005033160

Manufactured in the United States of America
1 2 3 4 5 6 – DP – 12 11 10 09 08 07

Introduction

Kenya is located in East Africa. The country is home to more than 32 million people. It has an area of 224,960 square miles. Kenya is almost as large as the state of Texas.

The official languages in Kenya are Swahili (swah-HEE-lee) and English. Swahili is the most common language spoken in Africa. It is a mix of African languages, Arabic, and English.

1 moja (MO-jah)

There is only **one** Mount Kenya. The top of Mount Kenya is more than 17,000 feet high. It is the highest place in Kenya and the second-highest in Africa. The weather in Kenya is very hot. But the top of Mount Kenya is covered with snow and ice. The mountain is so high that the air around it stays cold.

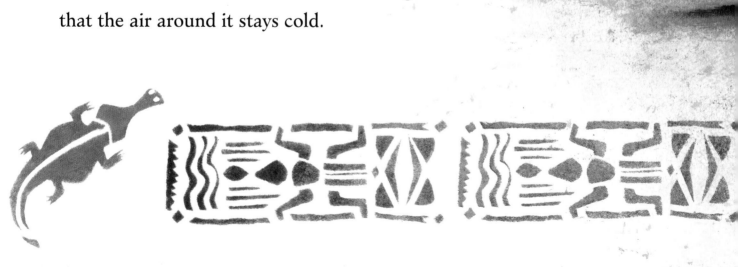

2 mbili (mm-BEE-lee)

Kenya's flag has **two** white spears. The color white stands for unity and peace. The spears are shaped like those used by a group of people called the Masai. These Kenyans are known as fierce warriors.

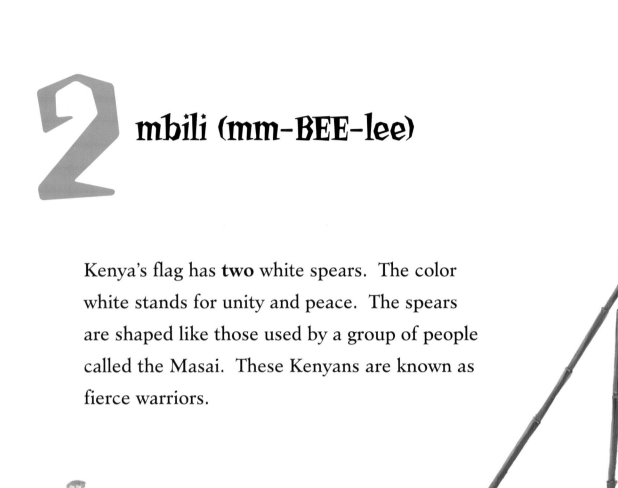

 tatu (TAH-too)

Three crafts in Kenya are soapstone carving, wood carving, and jewelry making. Soapstone is a soft rock. It is often carved to look like animals. Wood carvings may be almost any shape. Jewelry in Kenya is very colorful. It often includes beads and shells.

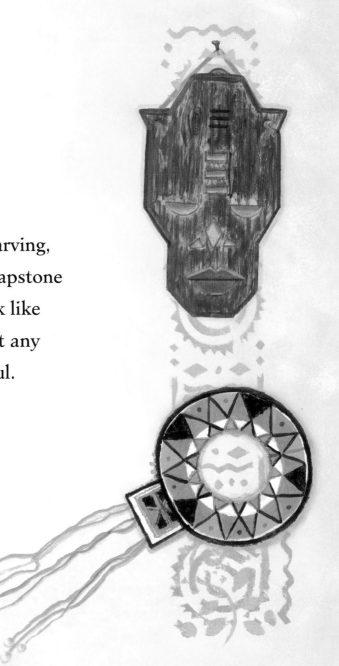

4 nne (NN-nay)

Four popular foods in Kenya are *ugali*, *uji*, chapati, and roasted corn. Ugali is a stiff dough that is eaten with stew. Uji is a porridge made from corn meal. Chapati is a flat bread. Kenyans eat roasted corn as a snack.

5 tano (TAH-no)

Five countries border Kenya. They are Ethiopia, Somalia, Tanzania, Uganda, and Sudan. Kenya is also next to an ocean and a lake. The Indian Ocean is on the southeast. Lake Victoria is between Kenya, Tanzania, and Uganda. Lake Victoria is the second-largest freshwater lake in the world.

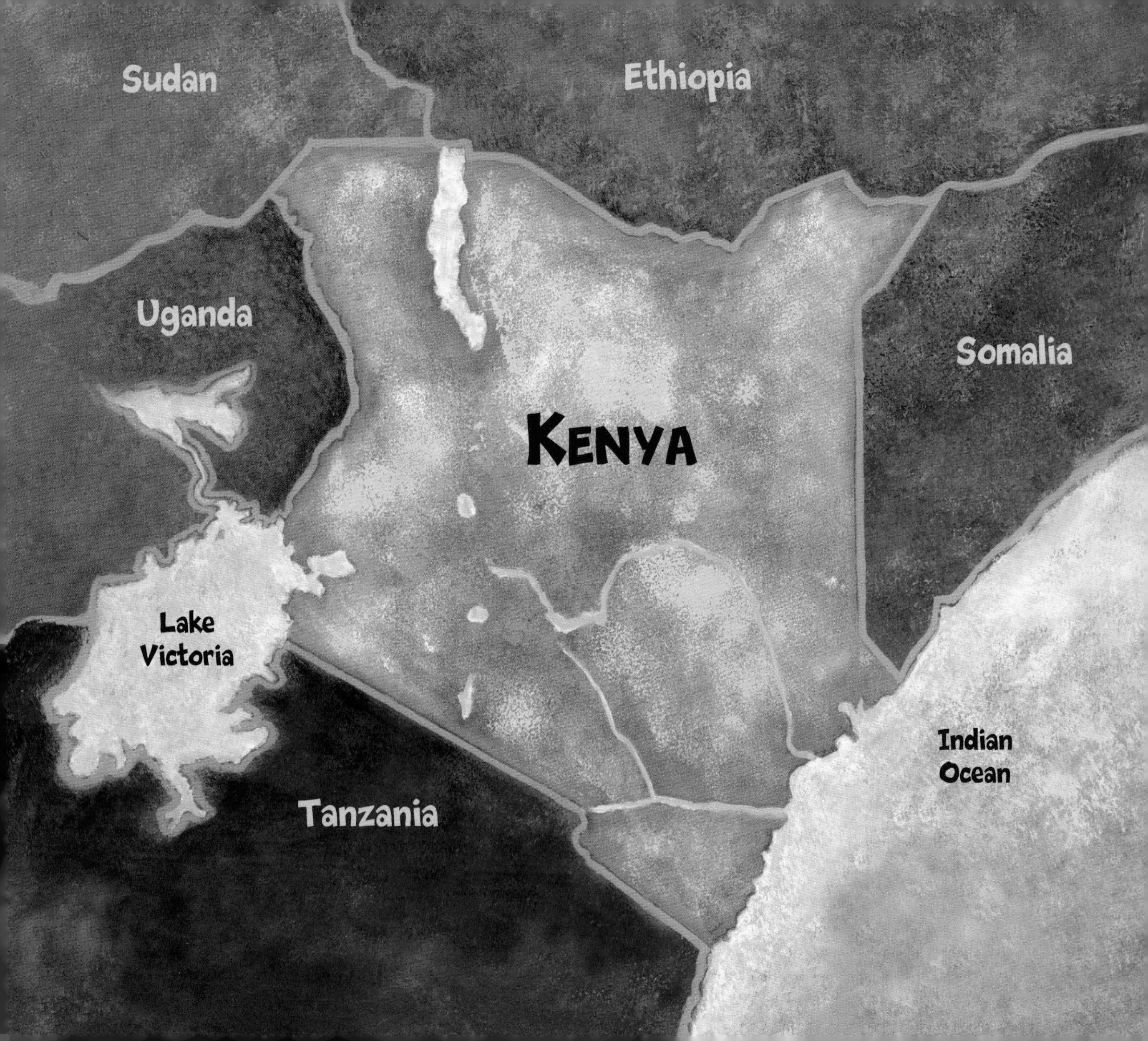
Sudan
Ethiopia
Uganda
Somalia
Kenya
Lake Victoria
Indian Ocean
Tanzania

6 sita (SEE-tah)

Six major crops in Kenya are bananas, corn, tea, sugarcane, coffee, and rice. Farming is very important to Kenya. Many people grow food for their families. Other people run large farms and sell their crops. More than half of all the goods that Kenya sells to other countries come from farming.

Bananas

Corn   **Tea**   **Sugarcane**   **Coffee**   **Rice**

7 saba (SAH-bah)

A child herds **seven** goats. Not all children in Kenya are able to go to school. Some stay home to help with chores. Taking care of a family's animals is a big job. People can use goats for milk or meat. They can also sell goats to make money.

8 nane (NAH-nay)

A musical instrument called a *nyatti* has **eight** strings. The nyatti is a traditional instrument of the Luo people. The Luo are one of many groups in Kenya. Other big groups include the Kikuyu, the Masai, and the Luhya.

9 tisa (TEE-sah)

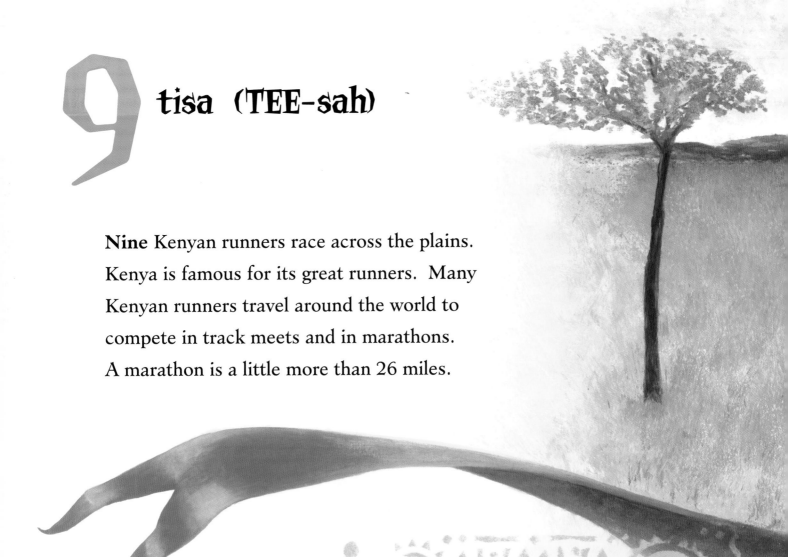

Nine Kenyan runners race across the plains. Kenya is famous for its great runners. Many Kenyan runners travel around the world to compete in track meets and in marathons. A marathon is a little more than 26 miles.

10 kumi (KOO-mee)

Ten wild animals that live in Kenya are buffalo, green pigeons, leopards, elephants, lions, baboons, zebras, ostriches, warthogs, and hippos. People from around the world come to Kenya to see its wildlife. They often go on trips called safaris to search for the animals.

Ostrich

Leopard

Buffalo

Green
Pigeon

Warthog

Hippos

Elephants

Baboons

Zebra

Lion

Pronunciation Guide

1 / moja (MO-jah)

2 / mbili (mm-BEE-lee)

3 / tatu (TAH-too)

4 / nne (NN-nay)

5 / tano (TAH-no)

6 / sita (SEE-tah)

7 / saba (SAH-bah)

8 / nane (NAH-nay)

9 / tisa (TEE-sah)

10 / kumi (KOO-mee)